Nuclear power:

Accidental releases — practical guidance
for public health action

Cover design: Michael J. Suess
Cover photo: Chernobyl nuclear reactor
near Kiev, USSR after the
accident on 26 April 1986.
Courtesy of POLFOTO, Copenhagen

World Health Organization
Regional Office for Europe
Copenhagen

Nuclear power:

Accidental releases — practical guidance for public health action

Report on a WHO meeting
Mol, Belgium, 1–4 October 1985

Nuclear power : accidental releases — practical
guidance for public health action : report on a WHO
meeting, Mol, Belgium, 1–4 October 1985
Copenhagen : WHO. Regional Office for Europe, 1986
WHO Regional Publications. European series ; No.21
ISBN 92-890-1112-2
    Nuclear Energy — Accident Prevention —
    Radioactivity — Accidents — Public Health —
    Congresses

ISBN 92 890 1112 2
ISSN 0378-2255

PRINTED IN ENGLAND

CONTENTS

NOTE

WHO policy in respect of terminology is to follow the official recommendations of authoritative international bodies, and this publication complies with such recommendations.

Nearly all international scientific bodies have now recommended the use of the SI units (*Système international d'unités*) developed by the Conférence générale des poids et mesures (CGPM),[a] and the use of these units was endorsed by the Thirtieth World Health Assembly in 1977. The following table shows three SI-derived units used frequently in this report, together with their symbols, the corresponding non-SI units and the conversion factors.

Quantity	SI unit and symbol	Non-SI unit	Conversion factor
Radioactivity	becquerel, Bq	curie, Ci	$1\,Ci = 3.7 \times 10^{10}\,Bq$ (37 GBq)
Absorbed dose	gray, Gy	rad	$1\,rad = 0.01\,Gy$
Dose equivalent	sievert, Sv	rem	$1\,rem = 0.01\,Sv$

---

[a] An authoritative account of the SI system entitled *The SI for the health professions* has been prepared by the World Health Organization and is available through booksellers, from WHO sales agents, or direct from Distribution and Sales, World Health Organization, 1211 Geneva 27, Switzerland.

Foreword

The WHO Regional Office for Europe has, over the years, developed a series of publications on public health aspects of nuclear power production and the disposal of radioactive waste. Following a report on the principles of public health action for accidental releases (1984), a working group was convened in Mol, Belgium in October 1985, to provide more detailed practical guidance in relation to such accidents.

The disaster at Chernobyl, USSR in April 1986 has dramatically highlighted the need for comprehensive contingency planning for — and emergency response to — such accidents. Such planning must be multisectoral and well coordinated, covering a wide range of governmental action at central, regional and local levels. Not least, the importance of clear guidance to the public must be recognized. There is also a need for close collaboration among neighbouring countries.

It must be accepted that there has been a major crisis of confidence in nuclear safety and, for its part, WHO is now embarking on an expanded programme related to public health aspects of radiation protection, to ensure that experience gained following the recent accident is fully evaluated and utilized.

Although the present volume was prepared before the Chernobyl accident, its conclusions are still considered as valid guidance to those concerned with the various aspects of emergency response.

It is appropriate that grateful thanks be recorded for the support provided by the Government of Belgium to this series of publications.

J.I. Waddington

Director, Environmental Health
WHO Regional Office for Europe

Introduction

In 1984 the WHO Regional Office for Europe published a report *(1)* on the principles of public health action in the event of the accidental release of radioactive materials into the environment, which is concerned largely with power reactors. Even though such reactors represent the major source of potential accidental release in many countries, they are by no means the only source, and emergency plans should also be developed for other types of nuclear installation. In addition, there may be unplanned releases from weapons accidents or terrorist activities which, by their very nature, defy the effective implementation of anything but the most general emergency plans. Nevertheless, the experience of planning to deal with accidental releases from nuclear installations will serve as a basis for the public health authority to deal with these other types of emergency.

In the previous report it was shown that accidents can be divided into three successive phases — early, intermediate and late — and that in each phase different decisions are required for action to protect members of the public. There is no single accident sequence that can be used for the preparation of emergency plans for different types of nuclear installation. However, the three phases appear to be common to all accidents and provide a framework within which radiological criteria can be established.

The previous report discussed the pathways of exposure relevant to each phase, and the potential consequences for health. In each phase the organs and tissues at risk were identified and both the stochastic and non-stochastic effects resulting from irradiation were considered and numerical risk values presented. In an accident that may lead to exposure of the public, the only means by which that exposure may be avoided or limited involve the introduction of countermeasures which interfere, to a greater or lesser extent, with normal living conditions. Countermeasures were identified that were applicable in each phase, and the risks and difficulties associated with introducing each countermeasure were discussed.

The principles evolved by WHO for protecting the public in accidents are consistent with the latest recommendations of the International Commission on Radiological Protection (ICRP) *(2)*. In the report of an ICRP task group *(3)* on the radiological criteria used for planning protection of the public in accidents, the principles established are as follows.

1

— Non-stochastic effects should be avoided by the introduction of countermeasures to keep individual doses below the thresholds for these effects.

— Individual risk from stochastic effects should be limited by introducing countermeasures that achieve a positive net benefit to the individuals involved. This can be accomplished by comparing the reduction in individual dose (and therefore risk) that would follow the introduction of a countermeasure with the increase in individual risk resulting from the introduction of that countermeasure.

— The incidence of stochastic effects should be limited by reducing the residual health detriment. This source-related assessment may be carried out by cost–benefit techniques and would be similar to a process of optimization, in that the cost of health detriment in the affected population is balanced against the cost of further countermeasures.

Following these principles, reference levels were proposed for the doses at which introduction of each countermeasure was warranted in each phase.

The present report is concerned with the implementation of the principles of the previous report. It is based on the collective knowledge and experience of the members of a Working Group, convened by the WHO Regional Office for Europe and WHO headquarters in collaboration with the Government of Belgium in Mol on 1–4 October 1985, to discuss practical guidance for public health action in the event of an accidental release of radioactive material. The Group included experts from 11 countries and representatives from international organizations such as the Commission of the European Communities (CEC), the Nuclear Energy Agency of the Organisation for Economic Co-operation and Development, and the International Commission on Radiological Protection.

Dr N. Wald was elected Chairman, Dr R.H. Clarke and Dr. J.-C. Nénot were the Rapporteurs, and Dr M.J. Suess and Dr P.J. Waight acted as Scientific Secretaries. The composition of subgroups formed during the meeting and the list of participants are given in Annexes 1 and 2, respectively.

This report is addressed to those organizations and individuals responsible for public health in the event of a nuclear accident. It will also be of use to those medical practitioners who are not administratively responsible in an accident, but who may need to be aware of the consequences and of action to be taken in the aftermath of an accident. Other organizations with direct responsibilities in the event of an accident will also need to become aware of the role and responsibility of the public health administration.

The guidance will be applicable to all types of nuclear installation and to accidental releases to both the atmosphere and the hydrosphere. However, to assist public health authorities in defining the size and scope of the contribution that might be required from them in case of an emergency, some information is given on the typical probability and magnitude of accidental releases from large nuclear installations and on the associated

2

range of radiological consequences. The example given refers to a standard pressurized light water reactor of the type currently used for electricity generation. Many of the features will be common to other nuclear installations, such as other types of reactor or fuel reprocessing plants. Atmospheric releases of radioactive material are generally of more concern because the probability of occurrence is higher and the potential exists for higher doses to be received in the short term. Accidental releases to the aquatic environment are less likely to occur and, in general, are likely to involve lower exposure levels and some delay before exposure of the population occurs; therefore, it is expected that this type of accidental release will leave enough time to implement the appropriate protective measures. Nevertheless, the principles for protecting the public following accidental aquatic releases will be the same as for releases to the atmosphere.

This report first summarizes the range of accident sequences for which plans need to be prepared for protecting members of the public. The measures that can be taken are described and the levels of dose at which they should be considered are summarized. Guidance is given on routes of exposure and the monitoring procedures that are likely to be applied to assess levels of exposure. The report then considers the problems that will need to be considered by public health authorities and by medical practitioners who will become involved in the provision of clinical services. The administrative arrangements applicable are outlined and consideration is given to the information and training aspects of planning to which public health authorities should have an input.

1

Purposes of planning
for radiation accidents

Safety analyses of nuclear installations usually identify and describe a wide range of potential accident sequences leading to exposure of the public. The predicted frequency of these accidents usually decreases as the magnitude of the corresponding releases increases.

Emergency plans should be designed to deal with an appropriately wide range of possible accidents. However, it would be a misallocation of resources to prepare detailed emergency plans and procedures for dealing with hypothetical accidents having extremely low probability of occurrence, even if the associated potential consequences may be very high. Therefore, such worst-case scenarios will not be covered in this report, although their consideration may be useful for other applications such as the characterization of the overall risk associated with a nuclear installation, for the purposes of siting and risk assessment.

The choice of the upper end of the frequency range of potential accident sequences on which to base emergency plans should therefore be an appropriate compromise between the requirement to protect the potentially exposed population (which would imply consideration, for planning purposes, of severe accidents) and that of practicability of emergency countermeasures (which would imply avoidance of committing disproportionately large amounts of resources to cope with very unlikely events). This threshold is frequently associated with accident sequences (called "reference accidents" in this report) which historically have had frequencies of occurrence in the range 10^{-4}–10^{-3} per year. The more modern designs of nuclear reactor have lower frequencies for the same magnitude of release *(4)* but emergency plans still consider the same order of release for preparatory planning.

Characteristics of Releases from Nuclear Installations

The majority of accidents requiring an off-site response will involve at some stage the potential or actual release of radioactive materials to the atmosphere. As emphasized in the Introduction, this report is primarily concerned with the consequences of these atmospheric releases.

The probability of occurrence, magnitude and isotopic composition of an accidental release will vary depending on the type of nuclear facility and

5

the severity of the accident. In the preparation of emergency plans, different source terms are considered, each being defined by the quantities of the different radionuclides liable to be released, their physicochemical form, the amount of time available before the release commences, and the expected duration of the release. These time factors are very important and may be decisive in the selection of the most effective and practicable protective measures to reduce the potential health consequences to the public.

The interval between the recognition of the start of an accident sequence having the potential for off-site consequences and the emergence of radioactive material into the atmosphere is important. If it is very short, only limited off-site action may be feasible before the release actually starts; this is improbable at large nuclear facilities with elaborate safety systems. In most cases there will be a delay before the uncontrolled release occurs, which may vary from about half an hour to one day or more *(5)*.

The duration of release also has important off-site consequences and may last from a fraction of an hour to several days *(1,5)*. Within this period there may be irregular and unpredictable peaks in the release rate. During the course of prolonged releases, changes may occur in the meteorological conditions, such as atmospheric stability, wind direction and velocity, or the presence and degree of precipitation. All these factors may modify the concentration of the dispersed radionuclides. For example, a change in meteorological conditions may well decrease the concentration, thus reducing the individual doses received, but may lead to population groups becoming involved who were not identified in the earlier stages.

At the stage of decision-making on emergency planning, it is for national regulatory authorities to decide on the level of consequences and the probability of occurrence that they are prepared to adopt in the definition of the reference accident. They would generally require that there exists a significant discontinuity in the probability–consequence relationship, so that more severe but very unlikely accidents can be discounted for emergency planning purposes.

The data and value ranges presented in the examples result from a review of a number of safety assessments carried out by competent authorities in several countries *(4,6–11)*. Table 1 shows the orders of magnitude of the release of the groups of radionuclides which are most relevant in the case of a reference accident in a light-water-cooled reactor producing 1000 MW of electricity per year. The possible range of radiological consequences associated with these typical releases depends on the distance to the nearest population group and on meteorological conditions. Orders of magnitude of individual doses are shown in Table 2 for the more important exposure pathways during the early phase of a release, namely the whole-body dose by external irradiation due to exposure to the airborne plume, the dose to the thyroid of children by inhalation of radioiodine from the cloud, and the dose to the lung by inhalation of radioactive aerosols. These doses are given merely to offer public health authorities an idea of the order of magnitude of individual doses liable to arise in the event of a nuclear emergency severe enough to activate the emergency plan. They should *not* be regarded as definitive for a particular nuclear plant at a given location.

6

Table 1. Example of a range of atmospheric releases from a reference accident

Range of annual probabilities	Total activity release (Bq)	Activity (Bq) associated with:		
		noble gases	iodine	other radionuclides (Ru, Cs)
10^{-4}–10^{-3}	10^{16}–10^{17}	~10^{16}–10^{17}	10^{13}–10^{14}	10^{13}–10^{14}

Table 2. Example of a range of radiation exposures from a reference accident

Type of dose (Sv)	Distance from the point of release		
	1 km	3 km	10 km
Whole-body (external irradiation)	10^{-2}–10^{-1}	5×10^{-3}–5×10^{-2}	10^{-3}–10^{-2}
Thyroid (inhalation)	10^{-1}–1	10^{-2}–10^{-1}	10^{-3}–10^{-2}
Lung[a] (inhalation)	10^{-2}–1	10^{-3}–10^{-1}	10^{-4}–10^{-2}

[a] The lung dose values depend heavily on the radioisotopic composition of the "other" nuclides released.

Source: **Kelly, G.N. et al.** *(12)* and **Charles, D. & Kelly, G.N.** *(13)*.

Time Phases

For the purposes of developing intervention levels three phases of an accident have been identified, which are generally accepted as being common to all accident sequences *(1,3)* — the early, intermediate and late (or recovery) phases. Although these phases cannot be represented by precise periods, and may overlap, they provide a useful framework within which the radiological criteria were established in the last report *(1)*.

Early phase
The early phase is defined by the period when there is the threat of a serious release, i.e. from the time when the potential for off-site exposure is recognized to the first few hours after the beginning of a release, if a release occurs. The interval between the recognition of an accident sequence and the start of

the release can be from less than half an hour to about a day *(1,5)* and the duration of the release may be between half an hour and several days. This variation in timing renders difficult decisions about the introduction of countermeasures, since there will be a need to forecast the future course of the accident and thus to predict doses and potential reductions of dose for situations that will not have arisen.

The feature common to both the warning period and the first few hours of release is that operational decisions are based on analysis of data from the nuclear installation itself and existing meteorological conditions. Thus decisions to implement countermeasures during the early phase will be based primarily on plant conditions and the associated potential doses to individuals in the population, assessed on the basis of prior analysis of plant fault sequences and probable meteorological patterns.

Some environmental measurements of off-site exposure rates and airborne concentrations from the plume may become available in this phase. Because of potential changes in release rate, meteorological conditions and wind direction, and in other unknown factors such as duration of release and the degree to which measurements represent future plume configurations, such measurements will be of minimal value for calculating projected doses.

Intermediate phase
The intermediate phase covers the period from the first few hours after the start of the release to one or more days. It is assumed that the majority of the release will have occurred at the beginning of this phase and significant amounts of radioactive material may already have been deposited on the ground, unless the release consisted only of noble gases. As previously stated, there is no clear boundary in emergency planning between the first and second phases.

It is during the intermediate phase that measurements of radioactivity in food, water and air, as well as radiation levels from deposited radioactive materials, will become available. The radiological characteristics of the deposited material will also be determined. Based on these data, dose projections can be made for principal exposure pathways, and these doses compared to pre-established intervention levels, so that decisions on the implementation of countermeasures can be made.

The intermediate phase ends when all the countermeasures based on environmental measurements have been implemented. If the accident is severe, the phase may be prolonged while extra measurements are made at locations further from the plant.

During the intermediate phase it would be expected that a group of experts would be formed from representatives of both the local and national authorities to advise on radiological protection of the public *(3)*. The responsibility for deciding on countermeasures involving the public may, in this phase, transfer from the operator who had such responsibility in the early phase to a government representative, who would be advised by the experts.

Recovery phase

The late or recovery phase is concerned with the return to normal living conditions. It may extend from some weeks to several years after the accident, the duration depending on the nature and magnitude of the release. During this phase the data obtained from environmental monitoring can be used to make the decision to return to normal living conditions, by the simultaneous or successive lifting of the various countermeasures decided during the first two phases of the accident. Alternatively, the decision could also be made to continue certain restrictions for long periods of time, with consequences for such aspects as agricultural production, occupation of certain areas or buildings, and the consumption of certain foodstuffs.

The withdrawal of countermeasures in the recovery phase will be based on analyses of actual cost, risk, benefit and societal impact of any residual contamination following decontamination, natural decay and weathering, and thus no predetermined levels have been provided for the withdrawal of countermeasures.

Health Effects

The previous report *(1)* described in detail the non-stochastic and stochastic effects, and only a brief summary is given here. The difference between non-stochastic and stochastic effects is illustrated in Fig. 1, which is based on an ICRP task group report *(14)*. Non-stochastic effects in individuals usually become more severe with increasing dose. In populations, an increase in dose may also result in increased frequency. Since the mechanisms of non-stochastic effects include cell death, and other effects may in themselves be observable at incipient stages, delineation of the dose–response relationship for any given type of non-stochastic effect depends on the stage and severity at which the effect is recognized. Fig. 1 shows how the frequency and severity of a non-stochastic effect, defined as a pathological condition, increase as a function of dose in a population of individuals of varying susceptibilities. The severity of the effect increases most steeply in those who are of greatest susceptibility (curve a), reaching the threshold of clinical detectability at a lower dose than in less susceptible subgroups (curves b and c). The range of doses over which the different subgroups cross the same threshold of detectability is reflected in the upper curve, which shows the frequency of the pathological condition in the population, and which reaches 100% only at that dose which is sufficient to exceed the defined threshold of severity in all members of the population.

For stochastic effects, as also illustrated in Fig. 1, the severity of the effect is independent of dose, and only the predicted frequency of the effect increases with increasing dose, without threshold.

Non-stochastic effects

The main interest in emergency planning is the identification of the dose levels below which non-stochastic effects are not likely to occur in a normal population. Non-stochastic effects can be induced in any organ or tissue given high enough doses. The discussion here is limited to effects in those

9

Fig. 1. Characteristic differences in dose–effect relationship between non-stochastic and stochastic effects

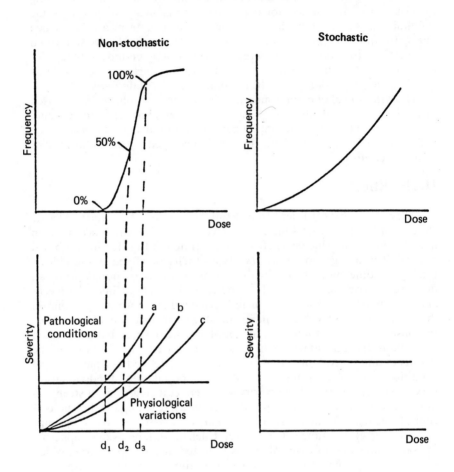

organs and tissues that are known to be most at risk from accidental releases from nuclear installations.

Whole-body irradiation at high enough doses will cause nausea, vomiting and diarrhoea; at even higher doses, early mortality will result from bone marrow cell depletion. Inhalation of large quantities of radioactive material will deliver high acute doses to the lung, leading to permanent impairment of lung function and even early mortality. Although severe irradiation of the gastrointestinal tract can also lead to early mortality, in nuclear accidents it is likely that irradiation of the bone marrow will be more important. Furthermore, at sufficiently high doses to the thyroid, non-stochastic effects may occur which may occasionally lead to death. Other non-lethal effects include impairment of fertility, skin damage and cataracts, but these are all less significant than those mentioned above. In addition, it should be emphasized that single-organ irradiation is most unlikely to occur in a nuclear accident, and that irradiation of several organs and tissues will be the most common type of exposure.

In the event of external irradiation *in utero*, the classic effects of sufficiently high doses on the developing fetus are gross congenital malformations, mental and growth retardation and death. For internal irradiation, differences in cellular metabolism may lead to different levels of risk. For example, the fetal thyroid is only at risk to ingested radioiodine when it is sufficiently developed to accumulate iodine.

Table 3 gives the levels of dose below which non-stochastic effects are not likely to occur in a normal population. It shows that, except for the fetus, the severe diseases and early deaths are related to high doses; accidents leading to such high doses will occur very infrequently.

Table 3. Levels of dose below which acute non-stochastic effects
are unlikely to occur in a normal population

Dose (Gy)	Organ	Effect
0.1	fetus	teratogenesis
0.5	whole body	vomiting
1	whole body	early death
3	gonads	sterility
3	skin	depilation, erythema
5	lens	cataract
5	lung	pneumonitis
10	lung	early death
10	thyroid	hypothyroidism

11

Stochastic effects

The stochastic effects following irradiation are either late somatic or genetic. The late somatic effect of primary concern is the increased incidence of fatal and non-fatal cancers in the irradiated population. The appearance of these cancers is usually delayed and may be spread over several decades. These late somatic effects include cancers for which the cure rate is low (lung, leukaemia) and others for which the cure rate is high (skin, thyroid). However, any cancer causes psychological effects that can significantly reduce the quality of life. There is a risk that serious hereditary disease may occur in subsequent generations following irrradiation of the gonads.

Risk factors for stochastic effects are given in the previous WHO report *(1)*. These risk factors are values averaged over all ages and for both sexes of a normal population. It should be recognized that these risk factors are based on the assumption of a linear dose–response curve without threshold, and consider only fatal cancers. For specific organs these factors may vary substantially with age, sex and other variables; consequently, they may lead to overestimation or underestimation of the risk. However, as they do not take into account the non-fatal cancers, such as thyroid and skin cancers, they may underestimate the total risk of cancers of some specific organs or tissues. These risk factors should therefore be considered as approximate values, and used as such.

Psychological effects

In addition to the predicted physical health consequences of irradiation, considerable psychological effects may constitute a significant public health problem. In contrast to the health effects previously described, the level of anxiety generated by possible exposure is not related to the level of exposure. Psychological stress may well be exhibited where radiation is low or insignificant. Psychological effects can be attributed to:

— the association of nuclear accidents with the explosion of a nuclear bomb;

— the inability of the human senses to detect ionizing radiation;

— inadequate and often conflicting information concerning the accident.

Recognition of this potential problem, and planning to deal with it, is an essential component of emergency preparedness.

Countermeasures and Objectives of Emergency Planning

Emergency measures designed to reduce adverse health effects are of two types: those that reduce the radiation exposure (protective measures) and those that reduce the health consequences of accidental exposure (medical care). The potential protective measures that could be implemented are:

— sheltering

— stable iodine administration

— control of access to the affected area

— evacuation

— relocation

— control of food and water supplies

— personal decontamination

— decontamination of areas.

The implementation of one or more of these measures depends not only on the nature of the accident, and its time phase, but also on specific local conditions such as population size and climatic and meteorological conditions. As a general principle, it is reasonable only to implement those protective measures whose social cost and risk will be less than that incurred by the radiation exposure.

Sheltering

A significant reduction in whole-body and skin doses due to external irradiation can be achieved by remaining indoors during the early phase. A substantial reduction in inhalation dose, affecting thyroid and lung, can also be achieved by closing windows, doors and other openings and switching off any ventilation systems. The shielding dose reduction factor provided by buildings can vary from 0.2–0.8 in the plume to 0.08–0.4 from deposition (1). Appropriate ventilation control can result in a reduction of inhalation dose by about a factor of 10.

The risk and harm resulting from short-term sheltering are low. Unplanned long-term sheltering can lead to social, medical and psychological problems.

Stable iodine administration

The administration of stable iodine compounds is effective in reducing the uptake of radioiodine by the thyroid gland. It is most effective when ingested prior to or at the time of exposure, and rapidly loses efficacy if administered a few hours after exposure. Consequently, it is necessary to ingest the stable iodine as soon as possible when a significant radioiodine release is predicted.

The recommended dosage of stable iodine compounds (KI or KIO_3) is 100 mg iodine equivalent daily for those over 1 year of age, and 50 mg iodine equivalent daily for infants. Undesirable but relatively minor side effects may occur in a very small proportion of people. In many circumstances it is unrealistic to attempt to distribute stable iodine to the population at risk once the accident has occurred; prior distribution is recommended either to individual dwellings or to focal points from which the iodine can be made available within a short time.

Control of access

Controlling the movement of people to and from the area affected by the accident will reduce the number exposed and facilitate emergency operations. Difficulties may arise if this countermeasure is maintained, as population groups may be anxious to move from or to return to their homes, to tend to domestic animals, or to salvage goods or products from the closed areas. With adequate control, the risk of traffic accidents should be minimal.

Evacuation
Evacuation is effective against external and internal exposure, but is a very disruptive measure and most difficult to implement. This is particularly true when large populations are involved. It should therefore be applied only when absolutely necessary to avoid short-term accumulation of doses leading to non-stochastic effects, and as far as possible to small population groups in the vicinity of the nuclear facility. It should be remembered that evacuation requires time to implement and will probably be most effective either if there is sufficient warning before a release or if it is used to avoid exposure to deposited radionuclides during the intermediate phase. Any emergency plan should take into account the private exodus of people from both affected and unaffected areas so as to minimize the disruptive effect. Although the social and economic costs of evacuation may be high, the risks to health are considered to be relatively small and will primarily result from traffic accidents.

Relocation
Relocation is implemented to avoid long-term high doses from the ground deposition of radionuclides, usually after the release has ended. It is less urgent than evacuation, and may be either short- or long-term. It is expensive, and depends on the availability of an appropriate reception area. The stress involved in relocation should not be underestimated.

Control of food and water supplies
Food control may entail destroying contaminated foodstuffs or restricting or banning their consumption, delaying their consumption by converting them to other products (e.g. milk to cheese), or storing them until the activity decreases to an acceptable level.

Control of water supplies usually means prohibiting the use of water from a contaminated source for any purpose.

Such measures may cause other problems in areas where there is already a shortage of food and/or water.

Personal decontamination
Personal decontamination should be undertaken only where there is evidence or a strong suspicion of body surface contamination. In general, domestic showers are adequate for decontaminating the skin, and most contamination of clothing can be removed by laundering.

Medical assistance may be required if there are contaminated injuries or where contamination cannot be removed by repeated washing. The only risk from personal decontamination is that of spreading radioactivity to previously uncontaminated areas.

14

Decontamination of areas

This protective measure involves the removal of contamination from the affected area to another location where it will be less hazardous. It may consist of washing, vacuum cleaning surfaces, ploughing agricultural land, or removing surface layers of soil. These measures are effective in reducing external radiation from deposited radioactivity and in restricting internal doses from the inhalation of resuspended radionuclides. The risk is to those who are exposed in performing the procedures.

The applicability of protective measures during various phases of the accident is shown in Table 4.

Table 4. Range of applicability of various countermeasures

Countermeasure	Phase		
	early	intermediate	late
Sheltering	+	±	—
Radioprotective prophylaxis	+	±	—
Respiratory protection	+	—	—
Body protection	±	±	—
Evacuation	+	+	—
Personal decontamination	±	±	±
Relocation	—	+	±
Control of access	±	+	±
Food control	—	+	+
Decontamination of areas	—	±	+

+ = Applicable and possibly essential.
± = Applicable.
— = Not applicable or of limited application.

Source: **International Atomic Energy Agency** *(5)*.

Guidance on Dose Value for the Introduction of Protective Measures

The principles for protection are identified in the Introduction to this book as:

— avoidance, if possible, of non-stochastic effects in individuals;

15

— limitation of individual stochastic risks by balancing the risk and cost of countermeasures against the risk and cost of further exposure;

— limiting the residual health detriment in the affected population.

The individual exposure should be as low as reasonably achievable, taking into account the risk of exposure and the risk associated with the countermeasures.

The decision to implement a protective measure, particularly in the early and intermediate phases, must be made on the basis of the risk to the potentially exposed individual, and the dosimetric quantity used must, therefore, express this risk. The quantity "effective dose equivalent" has been recommended (1, 15) for expressing the risk to individual members of the public during normal operation; this cannot, however, be applied to non-stochastic effects following accidents, since the risk coefficients and their associated weighting factors are based on fatal cancer incidence and serious hereditary defects with the assumption of proportionality between dose and risk. Therefore, the quantity that should be used to evaluate the non-stochastic effects will be the *absorbed dose*. In most accidents, the primary exposure of the public will be from beta- and gamma-radiation, and *dose equivalent* may be considered the suitable dosimetric quantity for expressing the stochastic risk to the individual.

Where an intake of radioactive material occurs at levels at which non-stochastic effects cannot occur, the individual *committed dose equivalent* is generally an accepted quantity to be applied to members of the general public. Other dosimetric quantities, such as the *collective dose*, will be of interest in decision-making during the late phase as part of an input to the general process of cost–benefit analysis.

Establishment of ranges of individual dose
It is clear that the risks, difficulties and disruption that follow the implementation of the various protective measures are widely different and thus the level of dose at which a given protective measure will be introduced is influenced by such considerations as well as by other site-specific factors. For these reasons, it is not possible to set one generally applicable intervention level at which a particular action would always be required. On the other hand it should be possible to define for each protective measure, on radiation protection grounds, a lower level of dose below which the introduction of the protective measure would not be warranted, and an upper level of dose for which its implementation should almost certainly be attempted. These two levels may be of guidance to national authorities when setting criteria for introducing protective measures.

The early phase
The introduction of sheltering for a limited period of time and, where appropriate, the administration of stable iodine, are countermeasures that have been accepted by many national authorities as constituting only a small risk to the individual. On radiological protection grounds the introduction

HOW TO RENEW ONLINE

- Select **My Account** from the library catalogue search screen.

- Enter your **Barcode Number** which is your student number including the card issue number i.e. 01 and **Password** – the first four letters of your last name.

- Click **Logon.**

- Click on the empty boxes next to the titles that you wish to renew and then select **Renew Loan.**

- The next screen will list your **Successfully Renewed Items.**

Online renewals can be made 3 times for a one week loan and 5 times for a one day loan before you will need to return the item for re-issuing.

PROBLEMS? You will not be able to renew online if you have outstanding fines, overdue books or messages on your record. Please see a member of the library staff or phone **Llandaff Learning Centre** on **029 2041 6244** for assistance.

<u>http://www.uwic.ac.uk/library/</u>

HOW TO RESERVE AN ITEM

- Search for the item on the library catalogue. Once found, click the small grey box labelled **"request"**.

- Enter your **Barcode Number** and **Password** (see over for more info).

- Choose to **Place a reservation** (**'Place a booking'** refers to four-hour loan items at Cyncoed and Howard Gardens only).

- Click **Place request.**

- **Choose your collection site** and click **Confirm.**

- You should now have successfully reserved an item.

Reservations should only be placed on items that are on loan to another borrower or on the shelf at a different site. You are encouraged to collect items currently available on the shelf at Llandaff yourself.

of such countermeasures would not appear to be warranted at projected doses, liable to be received in the short term, that are below the dose limits recommended for members of the public in any one year (5 mSv). It would seem reasonable that the levels of dose at which these countermeasures would almost certainly be justified be set an order of magnitude higher.

Evacuation is the most disruptive of the countermeasures that have been identified as applicable in this phase. Consideration of its introduction should start at dose levels significantly higher than those for the counter-measures mentioned above. Although it is difficult to justify choice of a particular value, the level of projected dose liable to be received in the short term, below which evacuation would not be justified, is likely to be about an order of magnitude greater than the dose limits for members of the public in any one year. The overriding aim in introducing countermeasures in the early phase is avoidance of non-stochastic effects. Therefore, evacuation should certainly be undertaken if the projected doses are liable to exceed those above which non-stochastic effects may occur. The resulting most restrictive upper and lower dose levels for the most common protective measures applicable in the early phase are shown in Table 5.

Table 5. Dose levels for early-phase protective measures
as developed by ICRP

Protective measure	Dose (mGy)	
	Whole body	Lung,[a] thyroid and any single organ preferentially irradiated
Sheltering and stable iodine administration	5–50	50–500
Evacuation	50–500	500–5000

[a] In the event of high-dose alpha-irradiation of the lung, the numerical values of the absorbed dose will be multiplied by a factor of 10, reflecting the relative biological effectiveness.

Source: **International Commission on Radiological Protection** *(3).*

The intermediate phase
The additional countermeasures applicable in the intermediate phase include restricting the distribution and consumption of locally produced water and fresh food and relocating groups of people pending decontamination of land or buildings. The disruption associated with countermeasures involving controlling food and water may be much less than that associated with relocation, which would be likely to be introduced to avert a higher level of projected dose. In general, there should be little penalty in not distributing fresh food, including milk. It may be appropriate to control the

distribution and consumption of fresh foods if the projected committed dose equivalent within the first year would otherwise exceed the dose limit for members of the public in any one year. However, under certain conditions, such as the unavailability of alternative supplies, it may be appropriate to allow a higher level of dose. The dose levels at which relocation would be considered depend largely on the size of the population affected.

When defining radiological criteria, one may consider that the annual dose equivalent limits for members of the public are clearly set at a low level of risk; the levels at which relocation would be considered should be significantly higher, and a factor of 10 seems appropriate. The time over which the contamination persists will affect decision-making; for example, it may be acceptable to allow people to receive higher doses in the first year after an accident if the annual projected dose is expected to decrease rapidly. In addition, the national interest may dictate that an industrial activity be continued in a contaminated area where the dose to essential personnel exceeds the annual occupational dose limit (50 mSv).

For both control of foodstuffs and relocation of population groups, the level of dose at which these protective measures should certainly be implemented should be an order of magnitude greater than the levels suggested for considering their possible introduction.

The resulting upper and lower dose levels for protective measures applicable in the intermediate phase are shown in Table 6. As with the protective measures applicable in the early phase, national authorities should give special consideration to the implications of irradiation of pregnant women and other special groups.

Table 6. Doses for intermediate-phase protective measures
as developed by ICRP

Protective measure	Dose (mSv or mGy) committed in the first year	
	Whole body	Individual organs preferentially irradiated
Control of foodstuffs and water	5–50	50–500
Relocation	50–500	not anticipated

Source: **International Commission on Radiological Protection** (3).

It will need to be decided at the time of an accident whether or not to implement an appropriate protective measure. This decision will be influenced by many factors involving the actual or potential release and the prevailing environmental and other conditions.

18

The above principles form the basis on which appropriate national authorities can specify levels at which emergency action would be implemented. In some cases all the quantitative data necessary to determine the balance of risk may not be available. Under these circumstances some general guidance on dose levels for the implementation of protective measures may be useful. Because of differences between various sites and countries the particular levels may vary; it can easily be proved in specific cases that a risk–benefit analysis could lead to other values for the introduction of any given countermeasure.

The recovery phase
As indicated previously it is neither feasible nor necessary to provide predetermined dose levels for the withdrawal of protective measures in the late phase, since this will be based on analyses of actual cost/risk, of the residual contamination, and of the benefit to and impact on society of the maintenance of the protective measures introduced.

2

Assessment of off-site radiation exposure

During the early and intermediate phases of an accident there is a need to predict the doses of radiation that members of the public are likely to receive. The main difference between the early and intermediate phases is that in the early phase decisions are made on the basis of predictions of potential doses yet to arise, whereas in the intermediate phase decisions can be based on the results of the confirmed environmental monitoring.

Activity Distribution in the Environment and Relevant Exposure Pathways

During the early phase of the accident estimates of the potential doses to members of the public in the immediate locality are estimated from the expected or already measured values of activity release. A comprehensive emission monitoring system is required to enable an estimate to be made of releases from all critical leakage points. Predictions of future release rates must be based on the available information on the conditions in the plant and possible future developments. In some installations data from static (fixed post) monitoring devices in the environment may be available.

Fig. 2 illustrates the pathways by which material released to the atmosphere is dispersed throughout the environment, leading to human exposure *(16)*.

In the case of releases into the atmosphere, the most important pathways during the early phase are:

— external doses of beta- and gamma-radiation from airborne radioactive materials (external submersion dose);

— external doses of beta- and gamma-radiation from radionuclides deposited on the ground;

— external doses of beta- and gamma-radiation from contaminated clothes;

— the internal dose from inhaled radionuclides.

22

Fig. 2. Pathways of radioactive contamination of the environment following atmospheric release

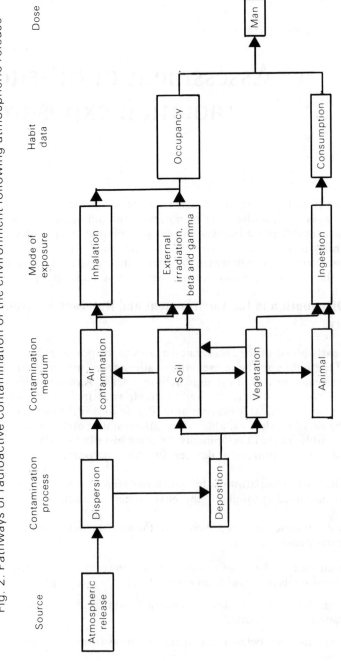

| Source | Contamination process | Contamination medium | Mode of exposure | Habit data | Dose |

Source: **International Atomic Energy Agency** *(14).*

It should be emphasized that rain showers over residential areas may lead to greatly increased local external exposure from deposited activity. Normally the contribution from deposited radioactive material that is re-suspended in the atmosphere will be small and may be neglected in the early phase.

Because of the delay in the distribution of foodstuffs and water, it is not usually necessary to ban their consumption in the early phase. Such decisions do have to be taken, however, in the intermediate phase. Ingestion can lead to exposure of people outside the contaminated area if no measures are taken to stop the distribution and consumption of foodstuffs produced in the contaminated area.

In the event of accidental releases into surface water the most important exposure pathways are:

— use of contaminated water for drinking and cooking;

— ingestion of contaminated fish;

— irrigation of plants with contaminated surface water.

In general, exposure from accidental releases into surface water can be reduced more efficiently than that from releases into the atmosphere.

Dose Estimation

Table 7 outlines the sequence of steps in evaluating the potential dose from accidental releases into the air.

The need in the early phase is to predict doses over short distances from the installation, typically of the order of a few kilometres. This is because decisions on the introduction of countermeasures are based on levels of dose to individuals. The first problem is that of the likely total quantities of radionuclides to be released. This can be estimated from knowledge of the condition of the installation, but a valuable source of information will be the stack monitoring information, if containment is not breached. Such monitoring equipment should be installed and should be capable of response in the event of a release several orders of magnitude greater than that anticipated in normal operation. In the early phase it is important that simple models for activity distribution in the environment and dose evaluation are applied. This is justified taking into account the uncertainties in the release predictions and the limited data available from environmental monitoring.

For evaluating the activity concentration in air as a function of distance, simple gaussian dispersion models will be sufficient in many cases, taking into account the meteorological conditions (wind direction and velocity; atmospheric stability) during the release period. For activity reaching the ground by wet and dry deposition, and for transfer through food chains, models have been developed that take into account the influence of seasonal variation. Detailed description of these radioecological models is outside the scope of this report; the reader is referred to special reports on the subject published by international bodies (16–19) and national authorities (4,11,20–24).

23

Table 7. Evaluation of exposure from accidental releases of activity into the atmosphere

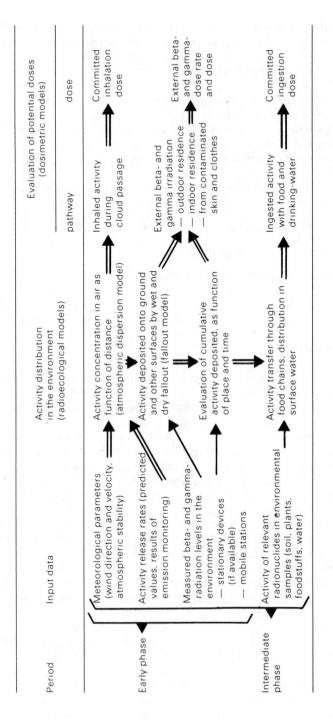

Although there are significant uncertainties involved in both the characterization and measurement of meteorological conditions, the uncertainties in source terms mean that simple models should be adequate in most cases. For some sites more sophisticated models may have been developed and these can be used if the detailed meteorological parameters are available at the site. This may require the linking of real-time site measurement data to national or regional meteorological forecasting services. In all of this prediction it is absolutely imperative that information on such factors as wind direction be unambiguous.

Although some monitoring results will become available in the early phase, there will be difficulty in using these results because of possible variations in release rate and meteorological conditions. In the intermediate phase, most of the release will have occurred and the primary requirement is to establish the extent and levels of the residual ground contamination. Initially, monitoring will be directed towards identifying higher levels of contamination in order to specify areas in which further countermeasures will need to be considered. However, it is important for monitoring to be undertaken well outside the areas where action might be anticipated. Such monitoring will ensure that competent authorities will be aware of all areas where there is measurable residual contamination.

For the evaluation of the dose to man from the different exposure pathways, the dosimetric models recommended by ICRP (25) can be applied. The derived dose coefficients, giving the dose per unit of inhaled or ingested activity, refer to adult members of the population. For dose estimation during the early phase and subsequent decisions on countermeasures, it may not be necessary to consider age-dependent dose coefficients, except in the case of estimating the dose to the thyroid, which is highly age-dependent (26).

However, in the intermediate and recovery phases, appropriate age-dependent dose data should be used to obtain more realistic estimates of the cumulative dose over both short and prolonged periods (estimation of lifetime dose).

It is generally true that intake of activity by inhalation occurs only during the period when people are actually immersed in the radioactive plume. In contrast to the inhalation dose, the external beta- and gamma-dose rate from deposited activity is proportional to the cumulated activity per unit area, taking into account appropriate shielding factors for those indoors at the time. After the end of the period of deposition, the external dose rate decreases according to radioactive decay and the "run off" of the deposited activity.

The assessment principles and techniques that can be used range from simple measurement and calculation methods using inexpensive equipment ("rules of thumb", formulae, charts or graphs) to sophisticated and complex computer techniques with colour visual display units. Prominence is often given to computerized methods, but emphasis on sophisticated computer-based technology should be regarded as an indication of its rapidly developing capabilities rather than as implying that the latest systems are essential for the task of assessment. Even a small microcomputer can make a large difference in the speed and reliability of an assessment calculation, which may have to be performed repeatedly for many hours.

The disadvantages of a totally computer-based assessment plan must be pointed out. As a back-up against failure of computer-based assessment schemes, it is recommended that a catalogue of pre-calculated geographic dose distributions from sample scenarios be prepared to define the scope of the problem and to direct further calculation more effectively. Furthermore, it is prudent to cross-check some important computer numerical results against manual calculations for order-of-magnitude correctness. This will require data on population densities, dose–intake relationships, and the composition of the population. It must be recognized by public health authorities that computers can create an illusion of infallibility, while in reality there is potential for large errors caused by mistakes in programming or data entry.

This report is concerned with assessing off-site consequences that can be expressed as (or are directly related to) radiation doses to individuals, in order to make correct decisions about countermeasures. It is not concerned with the assessment of other kinds of off-site consequences such as economic, social or political consequences of the accident. These may, however, have to be taken into account as factors in decision-making when, in a serious emergency, important decisions have to be taken by those who are probably non-technical but are at or near the political level. The report makes no attempt to indicate the weighting that should be given to factors of this kind.

Environmental Monitoring: Objectives and Procedures

As soon as the release has begun, it is possible to begin to monitor the levels of environmental radioactivity. This monitoring is important to complement the modelling in the early phase and is the basis for decision-making in the intermediate phase. Indeed, although the models allow the prediction of levels of exposure in large areas within a short time, the results predicted are based on hypothesis and on dynamic (source-term, meteorological) parameters. Consequently, there are uncertainties that must be assessed from field measurements.

The procedure used can take advantage of results of measurements made at fixed points; these results will confirm or modify the values theoretically predicted. Furthermore, it would be difficult to decide on important countermeasures implying serious socioeconomic consequences only on the basis of model predictions. For this reason, monitoring will aim at identifying the higher levels of contamination so as to specify areas in which countermeasures will need to be considered.

It must also be noted that the evaluation of the level of radioactive contamination recorded by the monitoring instruments can provide valuable information on the origin of the exposure (cloud and soil deposition).

Measurements that will need to be made in the intermediate phase will include external dose rates of beta- and gamma-radiation above the ground (these will need to be made according to a predetermined protocol specifying height above ground, etc.). Rapid assessment of the area affected will require either the use of airborne monitoring or hasty deployment of vehicles. For the resulting information to be comprehensible at the emergency centre, good communications must be established between the centre and each environmental monitoring team and the transmission of information must follow a fixed and predetermined pattern.

Samples of water, soil, plants and milk will be taken and passed to analytical laboratories where gamma-spectrometry and other methods of radiochemical analysis can be used to identify the nuclides that have been deposited. If various laboratories are involved, standard procedures of sampling and measurement must be defined. Since in most cases the analytical capacity of these radiochemical laboratories will be limited, priority in the order of analysis will have to be decided.

The last part of the monitoring will be undertaken later, during the recovery phase; residual levels in drinking-water and of ground contamination and activity in milk and other foodstuffs will be determined not only in the areas in which countermeasures were considered but also at distances well beyond those areas.

3

Guidance on planning for the medical and public health professions

It is the responsibility of the public health and medical professions to take an active part in the planning for response to a radiological emergency, with the aim of minimizing the health impact to the public. All actions to protect the public from consequences of a nuclear accident have this same objective and, therefore, the public health and medical professions should be represented on all planning and decision-making bodies.

It is recognized that the medical profession is accustomed to responding to a wide variety of emergency situations, with the obvious exception of those rare accidents that involve large releases of radioactive material. The purpose of this section is to provide information and guidance on special planning needs concerning the public health and medical professions for accidents involving such radioactive material.

Because of differences from country to country, this guidance cannot cover all the details required for developing radiological emergency response plans for the medical profession. Additional needs for planning in particular countries should be identified during exercises and/or drills, which are an important part of every radiological emergency response plan.

Two levels of planning are necessary: one for public health officials who must deal with general public health decisions, and one for medical personnel in the hospitals and field stations who must deal with the general public (22). The planning elements discussed below are intended to provide only a listing of the major problems for which public health and medical officials should develop response plans. Some of the planning elements are discussed in more detail in other parts of this book.

Planning Elements for Public Health Officials

Protective measures and their risks
The protective measures of concern are shelter, evacuation, administration of stable iodine and restrictions on the intake of food and water. Public health officers must be aware that these measures, which are intended to reduce radiation exposure and associated risks, may result in risks of other kinds to the populations involved (1,3,5).

In the early phase of an accident, decisions to provide shelter, to evacuate or to administer stable iodine are likely to be made rapidly (minutes to hours after recognition that an accident has occurred) and thus an evaluation of relative risks by public health officials will be impracticable. Therefore, these risks must be considered carefully during the planning phase, when intervention levels are selected to provide reasonable assurance of a net risk reduction. Special consideration should be given to those under medical care.

Potential radiation hazards
Because of the different types of nuclear installation that might be the source of an accidental release to the environment, public officials should be familiar with potential hazards associated with facilities in their area and should plan accordingly. Planning should include determining the information needs to support public health decisions, and should provide for channels for the timely dissemination of that information.

Training
As accidental irradiation of workers and the public is a rare event, it is not to be expected that doctors, nurses, paramedical staff and other health care professionals will have sufficient knowledge about the nature of radiation and its effects on the human body. For this reason, it is very important that all health workers receive education on the effects on health of exposure to ionizing radiation, and training in the proper and safe handling of patients contaminated with radioactive material in an emergency *(22)*. They must also learn to identify the signs of radiation sickness and their development with time, not least because this will enable them to identify those who have not been seriously irradiated, but who fear that they have.

In the training programme it should be emphasized that the most probable accident to prepare for is unlikely to involve any seriously irradiated people needing highly specialized care. Individuals seeking medical help and advice will probably not have received a dose that could cause a nonstochastic effect or even a dose of any significance. However, a large number of people who have not been exposed to radiation at all will present themselves to medical staff. They will basically require reassurance. The proper handling of the psychological problems associated with a nuclear accident is therefore of great importance and must be included in the training programme.

Nevertheless, a certain number of people may be contaminated and they will have to be decontaminated, probably at the evacuation centres where the majority of those contaminated can shower and change clothes. These people will have been identified by the procedures in the emergency plan. A few may have been heavily contaminated and also injured, so that hospital care is necessary. In the training it should be stressed that it is very unlikely that any patient will be so contaminated that he will in any way constitute a risk to the hospital staff giving the treatment, given that proper precautions are taken. Life-saving measures such as treatment for shock and severe injuries should never be postponed for the sake of very ambitious decontamination procedures.

The need for facilities and equipment
Since the probability is slight that accidents will occur that are serious enough to cause acute radiation injury to the general population, emergency planners are advised against purchasing expensive special equipment or facilities to be set aside solely for use in caring for patients contaminated with or exposed to radioactive materials (27).

Medical facilities in the vicinity of nuclear plants are usually equipped to care for any injured workers who have been contaminated or exposed to high doses of radiation. These facilities can also be used on those very rare occasions when a member of the general public who is similarly injured needs treatment. Lists of special supplies and equipment needed for caring for contaminated patients can be found elsewhere (28,29).

Public information
The general public will require information on a wide variety of topics, some of which deal with public health considerations. Pamphlets should be prepared for handing out to the public, and training programmes or seminars on the emergency plan may be presented. Public health officials should ensure that the services to be provided by the medical profession are covered accurately and adequately.

After an accident has occurred the public may be concerned about many subjects. Thought should be given to anticipating questions of a medical or health nature, as well as the appropriate answers or sources of answers to these questions and methods of communication with the public. Convincing and prompt answers by an authority on health matters will generally go a long way towards reducing anxiety.

Public health needs for population dose estimates
Public health officials may be called on to estimate the health consequences (immediate effects, long-term carcinogenic effects, birth defects and psychological effects) of an accident. Once the accident is under control, and at the time of the withdrawal of countermeasures, the authorities will be expected to account for the consequences of their protective action and also to account to the public through the various committees and boards of inquiry about the total risk to the public from the accident. The public health authorities should be prepared to estimate the insult to the population based on the best available radiological data.

Years after the accident, it can be expected that the people around the facility — both exposed and non-exposed — may attribute the inevitable cancers and genetic effects that arise in any population to the accident. Since a significant fraction of the whole population will get cancers from causes other than the accident, there is a likelihood that many improper claims will be made unless there are adequate data to estimate the probability that the accident caused the cancers. Public health authorities should ensure that the data necessary to make these decisions are included in the planning.

Planning Elements for Medical Practitioners

Operators of medical facilities in the vicinity of nuclear plants generally have emergency response plans and associated training programmes for dealing with the more frequently encountered conventional emergencies. If the hospital is situated in the immediate vicinity of the plant, the planners should be aware that if the accident is severe another hospital may be required in the event that the first hospital needs evacuating. This points to a need to explain the possible consequences of a radiation accident involving members of the public to the local medical staff. The emergency plans for such hospitals should be amended as necessary to include the special requirements that may be associated with a radiological emergency. The following are some key items that should be considered in developing those plans. Important elements in the training programme that should accompany the emergency plan have already been discussed.

Evacuation and shelter

The medical profession has a special responsibility when it comes to planning the evacuation of hospitals and nursing homes, as well as evacuation of the infirm and handicapped living in their own homes. The risk of moving these people must be carefully weighed against the risk of any irradiation if they remain in the shelter of their house at the time of the accident. The final decision on evacuating these people is probably best taken by the doctors attending them.

In the event that a decision is taken by the appropriate authority to evacuate the general public from a certain area, the designated evacuation centres should be attended by some medical staff. The purpose of the medical presence at these centres should mainly be to reassure members of the public and, if necessary, to decide who needs further medical attention. Medical and nursing staff at the centres are probably the best people to reassure those who are worried about a possible dose of radiation, rather than officials from the nuclear plant or from responsible government agencies.

Care and treatment of injured people who may be contaminated

Contaminated people who may be injured and in need of medical aid will be taken to the designated medical facility *(27)*. The facility must be prepared to treat these patients without disruption of its normal services. Treatment of injuries should be a first priority over that of decontamination. The facility will need to consider the appropriate procedures and contamination limits for radiation monitoring of patients, and methods and/or supplies for decontaminating people. Levels of decontamination should be established as a part of the plan, so that priorities can be established for decontamination. It should be emphasized that the number of people likely to be contaminated to the extent that it will pose a health risk is quite small; however, the number of people who are very slightly contaminated, or not contaminated at all, may be very large. These people will need reassurance, and thus an effective monitoring programme is needed at the evacuation centres or at other places away from medical facilities. If this is not done, there is the risk that these "self-diagnosed" people may jam the medical services and prevent those who are injured from obtaining the help they need.

Advice to people who return to live in contaminated areas
After the early phase of the emergency and evacuation, it is to be expected that there will be measurable, albeit low, levels of contamination in the areas to which people are allowed to return. Public health agencies will need to be prepared to explain to people living or working in this contaminated area about the residual radiation risks, so that they can make an informed decision about returning. Guidelines should be established to minimize the uptake of any residual contamination, and attention should be paid to local food crops, water supplies, and resuspension of deposited contamination.

Communication
The single most important aspect of emergency response is the communication system. Experience has shown that when any major accident occurs — not just those involving radioactivity — the normal communication system breaks down, and therefore a reliable, competent alternative system of communication will be needed and must be available.

The planning authorities must identify the sources of information they may need to respond properly during the emergency, and the methods of communication needed to obtain such information so as to reduce confusion. They must communicate with the population of the area to assure them that there are enough properly trained medical personnel to take care of their medical needs. They will also need to communicate with the public on procedures for obtaining such medical services.

Another important line of communication is to an advisory group of technical experts competent in the radiation field. These experts should be identified along with communication methods. A list should be prepared in advance so that they may be rapidly mobilized to participate in the decisions that need to be taken in an emergency.

4

Administrative training and public information

Small, unplanned releases from nuclear facilities are not especially unusual events. However, increased public awareness and rapid dissemination of information about these events by the news media creates a need for accurate and credible responses by government authorities at all levels.

Responsibility for assessment, management, countermeasures and recovery may rest with different branches of government depending on the organizational arrangements in each country. These responsibilities may be further subdivided between the relevant authorities at the national level and those at the state or provincial level. Therefore, any generalizations made about how they might function must be modified for the specific arrangement within each country. This may complicate communications between countries when a nuclear emergency occurs near a national border, and such situations must be considered in the planning.

In general, the course of an accident will be influenced by the decisions of the operators of the facility within minutes or hours after the event occurs. However, once the event is detected and reported, the nuclear regulatory authority or its counterpart will be the most likely government department to assess the consequences and the action being taken to bring the situation under control. In some countries there are emergency management (disaster control) organizations that would receive the initial reports. It should be recognized that easy access to reports and announcements may lead to the news media disseminating information about the accident to the public before the appropriate authorities are informed of the facts. This press information may often be incomplete or inaccurate, and this complicates the role of the authorities when they attempt to present more accurate information. There is a danger that this will result in the undermining of public confidence in the authorities, since there is an apparent tendency to believe the views of the news media.

In spite of the number of different authorities involved in emergency planning and implementation, it is imperative that one authority be given the leading role, and that that authority should be responsible for promulgating advice and information on action taken. The concept of a leading authority or individual is also important in communicating with the press.

This will enable the other authorities to deal with the technical decisions at hand and eliminate the chance of the press receiving conflicting information.

In the event of a serious accident in which countermeasures are likely to be considered, the public health authorities must be involved in the decision-making. Public confidence can best be restored if decisions on such matters as evacuation, sheltering and thyroid blocking are based on the advice of the responsible public health authority, since all these decisions are in themselves risky, and must be balanced against the risk from the projected radiation exposure. Details of the radiation release will generally be made available to the nuclear agency or the emergency management agency either by the facility management or by the local environmental or health agency. The public health agency must participate as a member of the team with these other organizations in assessing the consequences and suggesting corrective action. But since the final decision on corrective action has implications for public health, it should be the responsibility of the public health agency to endorse it.

The public health role in such countermeasures or evacuation must be emphasized, since the disruption of the lives of evacuees — some of whom may be pregnant, disabled or in hospital — will require that clinical services are available at evacuation or relocation centres. Should these people become contaminated during the course of evacuation, the clinical personnel must be prepared to deal with decontamination in addition to other duties involved in health care. Food, water and sanitation services will also be needed along the route and at relocation sites, and this is also a public health responsibility.

In a serious reactor accident, the airborne release may extend tens of kilometres outwards from the facility, and the population within that area could become mildly contaminated under certain circumstances. Thus, there is a need for medical personnel within that area to understand how to deal with radiation emergencies including contamination.

Public health authorities must also be prepared to work with food and agriculture authorities to determine the impact of releases on the food, milk and water down wind of the accident. This means that radiation protection guides must be considered before the accident and included in the emergency plan. These guides should ideally be coordinated with similar guides developed by authorities in adjacent countries, so that action will be consistent. The basis for these guides should be a public health understanding of risk.

During the late phase of the accident, a decision must be made to terminate the countermeasures, and again the public health authority is in the best position to justify, on behalf of the government, the basis and effects of this decision. The information and assessment needed to make this decision will come from the measurements and calculations provided by the facility and the relevant government organizations.

Throughout the recovery phase of the accident, there is the potential for additional exposure of the population, depending on the severity and nature of the damage to the facility. Discharges to the atmosphere or hydrosphere may occur, either inadvertently or deliberately as part of the recovery phase.

Deliberate discharge may be necessary, for example to reduce high pressure inside the reactor building, and should take into account all the relevant factors that will influence the levels of dose delivered to the population (time of day, meteorological conditions, etc.). Such discharges may cause public concern and resistance. If the public health authority is involved in planning for any deliberate discharge, and is informed about the magnitude and consequences, then there is a greater chance that public fear will be minimized. If these events are explained by the nuclear energy authority without public health input, there is a greater potential for public anxiety.

The psychological impact of an accident on the population must not be underestimated. Again, the public health authorities are in the best position to keep these fears in perspective and to provide credible information to the public.

As stated previously, it must be recognized that many of the people who eventually develop cancer will attribute it to the accident, regardless of whether they were exposed or not. Determining the validity of any such claim will depend on comprehensive data being obtained at the time of the accident. That would include not only the best available dose information, but information on where these people were and for how long, so that in the future, the probability of any particular cancer being caused by the accident may be assessed. The public health authorities should consider this eventuality and should collect the information during the accident to make possible future epidemiological assessment.

In addition to the respective roles of the government authorities during the various stages of an accident, the role of planning and training must be emphasized. Only through proper planning and field exercises will the problems such as communications be identified and corrected. The consequences for mental and physical health should be included in the planning, scenario development and field exercises. The public health authority should participate with the other authorities in all these stages.

Involvement of medical and health care personnel in these exercises is important, and the public health authority is in the best position to coordinate the role that they will play in the accident. Since these clinical personnel are most unlikely to have been trained in radiation monitoring and protection, there is a need to keep them informed and advised.

Communicating with the local population (particularly officials, the medical community and the news media) prior to an accident is an effective way to keep them informed about the quality of the emergency plan. Involving these people in the exercises or in the review of the plans will go a long way towards building confidence in the facility.

References

1. *Nuclear power: accidental releases — principles of public health action.*
   Copenhagen, WHO Regional Office for Europe, 1984 (WHO Regional
   Publications, European Series, No. 16).
2. **International Commission on Radiological Protection.** Recommendations of the ICRP. *Annals of the ICRP,* **1**(3), 1977 (ICRP Publication 26).
3. **International Commission on Radiological Protection.** Protection of the
   public in the event of major radiation accidents: principles for planning.
   Annals of the ICRP, **14**(2), 1984 (ICRP Publication 40).
4. **Kelly, G.N. & Clarke, R.H.** *An assessment of the radiological consequences of releases from degraded core accidents for the Sizewell PWR.*
   London, H.M. Stationery Office, 1982 (NRPB Report R137).
5. *Planning for off-site response to radiation accidents in nuclear facilities.*
   Vienna, International Atomic Energy Agency, 1981 (IAEA Safety Series
   No. 55).
6. *Reactor safety study.* Washington, DC, US Nuclear Regulatory Commission, 1975 (report WASH-1400).
7. **Evans, J.S. et al.** *Health effects model for nuclear power plant accident
   consequence analysis.* Washington, DC, US Nuclear Regulatory Commission, 1985 (report NUREG/CR-4214, SAND 85-7185).
8. **Iljin, L.A. et al.** [Assessment of radiation consequences of nuclear power
   plant accidents and the problems of population protection]. *In:* [*Radiation safety and protection of nuclear power plants*], Vol. 8. Moscow,
   Energoatomizdat, 1984, pp. 146–154 (in Russian).
9. *Manual of protective action guides and protective actions for nuclear
   incidents.* Washington, DC, US Environmental Protection Agency,
   1980.
10. **Bundesminister für Forschung und Technologie, ed.** *Deutsche Risikostudie
    Kernkraftwerke — eine Untersuchung zu dem durch Störfalle in Kernkraftwerken verursachten Risiko.* Cologne, Verlag TUV, 1979.
11. **Bundesminister des Innern, ed.** *Leitfaden für den fachlichen Berater der
    Katastrophenschutzleitung bei kerntechnischen Notfällen.* Cologne, Institute for Accident Research, 1984.

12. **Kelly, G.N. et al.** *The radiological consequences of a design basis accident for the Sizewell PWR.* London, H.M. Stationery Office, 1983 (NRPB Report M104).

13. **Charles, D. & Kelly, G.N.** *Matrix of dosimetric data conditional upon the release of UK1 in particular meteorological conditions.* London, H.M. Stationery Office, 1984 (NRPB Report M116).

14. **International Commission on Radiological Protection.** Nonstochastic effects of ionizing radiation. *Annals of the ICRP,* **14**(3), 1984 (ICRP Publication 41).

15. *Basic safety standards for radiation protection.* Vienna, International Atomic Energy Agency, 1982 (IAEA Safety Series No. 9).

16. *Generic models and parameters for assessing the environmental transfer of radionuclides from routine releases. Exposures of critical groups.* Vienna, International Atomic Energy Agency, 1982 (IAEA Safety Series No. 57).

17. *Methodology for evaluating the radiological consequences of radioactive effluents released in normal operations.* Brussels, Commission of the European Communities, 1979.

18. *Assessment of off-site consequences of an accident in a nuclear installation: techniques and decision making.* Vienna, International Atomic Energy Agency (in press).

19. **International Commission on Radiological Protection.** Assessment of doses to man. *Annals of the ICRP,* **2**(2), 1979 (ICRP Publication 29).

20. **Clarke, R.H. & Kelly, G.N.** *MARC — the NRPB methodology for assessing radiological consequences of accidental releases of activity.* London, H.M. Stationery Office, 1981 (NRPB Report R127).

21. **Le Grand, J. & Manesse, D.** *Modèle IPSN pour le calcul simplifié de la dispersion atmosphérique des rejets accidentels.* Fontenay-aux-Roses, Atomic Energy Commission, 1982 (report CEA-R-5170).

22. **Shleien, B.** *Preparedness and response in radiation accidents.* Washington, DC, US Department of Health and Human Services, Food and Drug Administration, 1982 (HMS Publication FDA 83-8211).

23. **Simmonds, J.R.** *The influences of season of the year on the predicted agricultural consequences of accidental releases of radionuclides to the atmosphere.* London, H.M. Stationery Office, 1985 (NRPB Report R178).

24. **Haywood, S.M. & Simmonds, J.R.** *Agricultural consequences of accidental releases of radionuclides for atmosphere. Sensitivity to the dose criteria for restricting food supplies.* London, H.M. Stationery Office, 1985 (NRPB Report M124).

25. **International Commission on Radiological Protection.** Limits for intakes of radionuclides by workers, Parts 1–3. *Annals of the ICRP,* **2**(3), 1979; **4**(3/4), 1980; **6**(2/3), 1981 and supplements.

26. **International Commission on Radiological Protection.** Statement from the 1983 Washington meeting of the ICRP. *Annals of the ICRP,* **14**(1), 1984.

27. *A guide to the hospital management of injuries arising from exposure to/or involving ionizing radiation.* Chicago, American Medical Association, 1984.

28. *Manual on early medical treatment of possible radiation injury.* Vienna, International Atomic Energy Agency, 1978 (IAEA Safety Series No. 47).
29. *Management of persons accidentally contaminated with radionuclides.* Washington, DC, National Council on Radiation Protection and Measurements, 1980 (NCRP Report No. 65).

Membership of subgroups

Subgroup 1 on source terms

Mr F. Diaz de la Cruz
Mr F. Luykx
Dr V.N. Lyscov
Dr O. Ilari (*leader*)

**Subgroup 2 on health effects
and countermeasures**

Dr A.K. Bhattacharjee
Dr L.B. Sztanyik (*leader*)
Dr P.J. Waight

**Subgroup 3 on evaluation of
off-site consequences**

Dr G. Fieuw
Dr W. Jacobi (*leader*)
Mr R. Kirchmann
Dr V.N. Lyscov

**Subgroup 4 on administrative, training
and public information
responsibilities**

Mr J.C. Villforth

**Group 5 on planning guidance
for public health and medical
professionals**

Dr Celia T. Anatolio
Mr J.E. Logsdon (*leader*)
Dr S.E. Olsson
Mr J.C. Villforth

Subgroup 6, composed of
Drs Clarke, Nénot, Suess and
Wald, acted as a steering
committee to review with the
other subgroups the development
and content of their respective
assignments, and to integrate
their activities.

Participants

Temporary advisers

Dr Celia T. Anatolio, Director, Radiological Health Service, Ministry of Health, Manila, Philippines

Dr A.K. Bhattacharjee, Deputy Director-General (Medical), Directorate-General of Health Services, Ministry of Health and Family Welfare, New Delhi, India

Dr R.H. Clarke, Secretary, National Radiological Protection Board, Chilton, Didcot, United Kingdom (*Co-Rapporteur*)

Mr F. Diaz de la Cruz, Technical Adviser, Nuclear Safety Council, Madrid, Spain

Mr G. Fieuw, Head, Department of Measurement and Radiation Control, Nuclear Energy Research Centre (CEN/SCK), Mol, Belgium

Dr W. Jacobi, Professor and Director, Institute of Radiation Protection, Society for Radiation and Environmental Research, Neuherberg, Federal Republic of Germany

Mr R. Kirchmann, Manager, Radiation Protection Programme, Nuclear Energy Research Centre (CEN/SCK), Mol, Belgium

Dr A. Lafontaine, Professor, University of Louvain, Brussels, Belgium

Mr J.E. Logsdon, Health Physicist, Office of Radiation Programs, US Environmental Protection Agency, Washington, DC, USA

Dr V.N. Lyscov, Associate Professor and Head, Biophysics Section, Moscow Engineering Physics Institute, USSR

Dr J.-C. Nénot, Chief, Radiation Health Services, Department of Health Protection, Institute of Protection and Nuclear Safety, Atomic Energy Commission (CEA), Fontenay-aux-Roses, France (*Co-Rapporteur*)

Dr S.E. Olsson, Professor, Radiation Protection Medicine, National Institute of Radiation Protection, Stockholm, Sweden

Dr L.B. Sztanyik, Director-General, National Research Institute of Radiobiology and Radiohygiene, Budapest, Hungary (*Vice-Chairman*)

Mr J.C. Villforth, Director, Center for Devices and Radiological Health, Food and Drug Administration, Department of Health and Human Services, Rockville, MD, USA

Dr N. Wald, Professor and Chairman, Department of Radiation Health, Graduate School of Public Health, University of Pittsburgh, PA, USA (*Chairman*)

Representatives of other organizations

Commission of the European Communities (CEC)

Mr F. Luykx, Head of Section, Health Protection and Public Health, Health and Safety Directorate, Luxembourg

International Commission on Radiological Protection (ICRP)

Dr W. Jacobi

Organisation for Economic Co-operation and Development (OECD)

Dr O. Ilari, Deputy Head, Division of Radiological Protection and Waste Management, Nuclear Energy Agency, Paris, France

World Health Organization

Regional Office for Europe

Dr M.J. Suess, Regional Officer for Environmental Health Hazards (*Co-Scientific Secretary*)

Headquarters

Dr. P.J. Waight, Scientist, Prevention of Environmental Pollution, Division of Environmental Health (*Co-Scientific Secretary*)